Andréa Bicca Noguez Martins et al

Tests for seed analysis: A scientific approach

Literature review

ScienciaScripts

Imprint

Any brand names and product names mentioned in this book are subject to trademark, brand or patent protection and are trademarks or registered trademarks of their respective holders. The use of brand names, product names, common names, trade names, product descriptions etc. even without a particular marking in this work is in no way to be construed to mean that such names may be regarded as unrestricted in respect of trademark and brand protection legislation and could thus be used by anyone.

Cover image: www.ingimage.com

This book is a translation from the original published under ISBN 978-620-6-76124-2.

Publisher:
Sciencia Scripts
is a trademark of
Dodo Books Indian Ocean Ltd. and OmniScriptum S.R.L publishing group

120 High Road, East Finchley, London, N2 9ED, United Kingdom
Str. Armeneasca 28/1, office 1, Chisinau MD-2012, Republic of Moldova, Europe
Printed at: see last page
ISBN: 978-620-8-14549-1

Editors/Authors

Andréa Bicca Noguez Martins

Andressa Alves Cassão

Andressa Meneses Prockt

Bruna Franco da Silva

Daniele Pacheco da Silva

Gabriele Padilha Schneider

Joice Fernanda Lübke Bonow

Luis Filipe Oliveira de Leon

Paula Cilene Machado Munhoz

Pierri Caldas Bauerman

Willian Siqueira da Silva

Ygor Mota Soca Machado

Seed quality can be understood as a set of attributes, characteristics or components that determine the seed's performance in the field.

In order to elucidate the different aspects of seed production, it is necessary to constantly update technologies. The use of high quality seeds with high production potential, tolerance to stresses, resistance to pathogens, physical, physiological and sanitary attributes, as well as seed processing and storage processes, stand out.

The available bibliography needs information on the development and results of research into seed production.

Therefore, a group of teachers and students from the Federal Institute of Education, Science and Technology of Rio Grande do Sul's Bagé Campus have joined forces to compile existing information on seed analysis. It is hoped that this publication will help to encourage studies on this subject.

With this in mind, this book presents a study of scientific approaches to seed quality analysis. Spread over five chapters, the study covers the main points relating to seed quality.

Great reading!

Paula Cilene Machado Munhoz

Andréa Bicca Noguez Martins

Summary

Chapter 1

Germination test

INTRODUCTION

Germination is characterized by the development of the essential structures of the embryo, demonstrating that it will give rise to normal plants. It is preferable to use laboratory analysis methods, as the variable environmental conditions of the field can lead to inconsistent results. In the laboratory, controlled and standardized conditions allow for more uniform, rapid and complete germination, thus guaranteeing reproducible and comparable results, which are essential for an accurate assessment of seed quality. Therefore, the aim of the germination test according to the RAS (Rules for Seed Analysis) manual is to determine the maximum germination potential of a batch of seeds, compare the quality of different batches and/or estimate their value for sowing in the field. In view of the above, this chapter aims to discuss and compile current scientific work on germination tests (Brasil, 2009).

MATERIAL AND METHODS

This study consists of a bibliographic review, characterized by the survey, analysis and description of scientific publications in a specific area of knowledge. The bibliographic research was carried out in the SciELO (Scientific Electronic Library Online) virtual library and Google Scholar. Using the descriptor "Seed germination test", care was taken to ensure the accuracy of the information presented, with the aim of describing the concepts and practices most commonly used by authors.

The database search took place in June 2024, covering books, magazines and scientific journals published between 2009 and 2021. After reading the articles, all the material was compared in a detailed literature review, with the aim of this review being to provide a wide range of current concepts and practices on the importance of seed germination testing.

RESULTS AND DISCUSSION

Basic principles are used to conduct the germination test on different types of seeds, such as the type of substrate, temperature, duration of the test, light requirements, but there are specific instructions for each species, as well as the availability of suitable equipment.

TYPES OF SUBSTRATE

The substrates commonly used are: Paper, sand and soil, however, there are alternative materials that have been tested to make maximum germination of the tested seeds possible. According to Silva *et al.* (2017), who tested the germination and vigor of *Parkia platycephala Benth.* seedlings in different substrates and temperatures, it was possible to verify that among the substrates tested (vermiculite, sand, coconut dust, sugarcane bagasse, tropstrato®, Germitest brand paper (RP) and blotting paper.) in a *Biochemical oxygen Demand* (B.O.D.) germinator, it was possible to verify that an alternating temperature of 25-35 °C combined with the vermiculite substrate is recommended for germination tests of this species.

However, the ideal substrate varies according to the species. In tests carried out by Leão et al. (2015), testing the viability of ipê amarelo seeds in four substrates (sand, vermiculite, paper towels and blotting paper), it was found that the blotting paper substrate at 30 °C is ideal for carrying out germination tests in laboratory conditions.

The substrate should be chosen according to the characteristics of each seed and the standards established by the Rules for Seed Analysis (RAS).

DURATION OF THE TEST

The physiological effects vary between different seed species, which results in different counting times in germination tests. Some seeds germinate quickly due to more efficient metabolic processes, while others have dormancy that must be overcome, leading to longer germination times. Factors such as shell hardness, the need for treatments such as scarification or stratification and the specific environmental conditions of temperature and humidity also have an influence on the rate and time of germination. These physiological factors of each species determine the duration required to achieve accurate results in germination tests (Brasil, 2009; Da Silva, 2020).

LIGHT AND TEMPERATURE REQUIREMENTS

Temperature and light are fundamental in germination tests, as the right temperature activates essential metabolic processes, while light, depending on the species, can stimulate or inhibit germination. A number of studies have explored the influence of light and temperature on seed germination. Ramos (2020) found that for *Mimosa bimucronata*, constant temperatures of 25-30 °C and an alternating temperature of 20-30 °C with white light were optimal. Similarly, Hilgert (2021) reported that the absence of light and high temperature promoted germination in *Carya illinoinensis*, while Fernandes (2021) observed that Sporobolus indicus required light for germination, with 20-30 °C being the most suitable temperature. This shows the particularities of each species in terms of the need for light and temperature for germination.

EQUIPMENT

The germinators used in Seed Analysis Laboratories can be classified into chamber and room germinators. The chamber germinator has insulated double walls to reduce temperature variations, with trays for the samples and controls for heating, cooling, humidity, photoperiod and thermo-period. Sometimes it is necessary to wrap the substrates in containers resistant to water vapor exchange. The room germinator, on the other hand, allows people to enter, with samples on shelves or trolleys. Fans prevent temperature stratification and humidifiers maintain high relative humidity. The combination of both allows constant or alternating temperatures to be used as required. Two types of counters are often used: perforated plates and vacuum counters. Whenever possible, seeds should be counted with these devices, as they facilitate both the operation and the random distribution of seeds over the substrate (Brasil, 2009).

FINAL CONSIDERATIONS

Given this information, seed quality and germination tests are essential to ensure that only viable, high-quality seeds are used, in order to guarantee efficient agricultural production. Future studies should focus on improving germination techniques, including technological advances and adapting to factors that can generate risks for agriculture and thus continue to guarantee sustainability and global food security.

BIBLIOGRAPHICAL REFERENCES

BRAZIL. MINISTRY OF AGRICULTURE AND AGRARIAN REFORM. National Secretariat for Agricultural Defense. National Plant Defense Department. **Rules for seed analysis Brasilia**, 2009.

DA SILVA, G. A.; *et al*. Environmental factors in seed germination and defense mechanisms to ensure its perpetuation. **Pesquisa, sociedade e desenvolvimento**, v. 9, n. 11, pág. e93491110524-e93491110524, 2020.

FERNANDES, T.; *et al*. Germination of razor grass (*Paspalum virgatum l.*) and capeta grass [(*Sporobolus indicus* (l.) r. br.)] as a function of temperature and light. **Revista ibero-americana de ciências ambientais**, v. 12, n. 12, p. 84-91, 2021.

HILGERT, M. A.; *et al*. Luminosity and temperature on the germination of pecan seeds. **Pesquisa agropecuária gaúcha**, v. 27, n. 1, p. 74-89, 2021.

LEÃO, N. V. M.; *et al.* biometry and diversity of temperatures and substrates for the viability of ipê amarelo seeds. 2015.

RAMOS, M. G. DE C.; *et al.* Effect of light and temperature on the germination of Mimosa bimucronata (dc) o. kuntze seeds. **Revista craibeiras de agroecologia**, v. 5, n. 1, p. e9399-e9399, 2020.

SILVA, R. B.; *et al.* Germination and vigor of *parkia platycephala benth.* seedlings in different substrates and temperatures. **Revista ciência agronômica**, v. 48, p. 142-150, 2017.

Chapter 2

Accelerated Ageing

INTRODUCTION

The accelerated ageing test originated with Crocker and Groves (1915), who, when studying the viability of seeds, observed that the deterioration and death of seeds during storage was caused by the coagulation of proteins, a process accelerated by heat. However, it was only in 1965 that Delouche defined the concept of artificial ageing, based on a two-term algebraic expression in which seeds subjected to high temperatures and relative humidity have their deterioration increased (França-Neto and Krzyzanowski, 2018).

In 1973, Delouche and Baskin proposed the accelerated ageing method to predict the storage potential of seed lots, indicating that the loss of germination capacity precedes the death of the seed. This process involves a series of biochemical variations resulting from the ageing of the seeds, resulting in a decrease in germination speed, seedling emergence and an increase in abnormal seedlings. The storage potential of a batch of seeds is determined by the pre-conservation history or the level of deterioration from field ripeness to the end of storage, suggesting that the longevity of seeds is determined by their initial physiological quality (França-Neto and Krzyzanowski, 2018).

Currently, due to the importance of good crop performance in field conditions, seed technology seeks to improve tests to assess the germination potential and vigor of seeds (França-Neto *et al.*, 2007). Among the various tests used for this purpose, the accelerated aging test stands out as one of the most widely used worldwide, especially for corn and soybean seeds (Dutra and Vieira, 2004).

The accelerated ageing test is a valuable technique used to assess seed vigor in various plant species (Corrêa, 2017). By simulating adverse environmental conditions for a controlled period, it allows batches with lower germination and storage potential to be identified, providing essential information for proper seed management and optimization of agricultural production systems (Rodrigues, 2007).

Temperature is one of the main factors influencing the rate at which seeds age. Increasing the temperature generally accelerates the chemical reactions that degrade the cell structures and nutrient reserves of the seeds. Humidity also plays a crucial role, as it can promote the growth of fungi and other forms of deterioration. UV radiation can damage the seeds' DNA, affecting their ability to germinate and produce vigorous plants (Câmara, 2016).

The length of time the seeds are exposed to stress is also crucial, in order to strike a balance between accelerating the ageing process and maintaining fidelity to the actual conditions in which the seeds are stored and used. A very short exposure time may not be enough to cause significant damage, while a very long time may lead to the death of the seeds (Rosseto *et al.*, 1995).

The success of the accelerated seed ageing test lies in the careful choice of stress variables and the precise definition of the exposure time. These variables must be carefully selected to replicate the actual storage and use conditions of the seeds and to accelerate the ageing process proportionally (Bertolin *et al.*, 2011).

In view of the above, the aim of this review was to compile an overview of the accelerated ageing test, in order to present the steps to be followed when carrying out the test correctly.

MATERIAL AND METHODS

This chapter was written in such a way as to provide a theoretical approach to the accelerated ageing test. To gather bibliographic information, documents and information were searched and consulted on digital academic research platforms such as Scielo, Google Scholar, as well as scientific websites and books.

RESULTS AND DISCUSSION

Seed ageing is a natural process influenced by various environmental factors such as temperature, humidity and oxygen. The test is based on subjecting seeds to conditions of environmental stress, such as high temperature and high relative humidity, for a defined period. During this period, seeds with lower vigor suffer more severe physiological damage, compromising their germination capacity and the vigor of subsequent seedlings. After accelerated ageing, the seeds are subjected to germination tests under ideal conditions, allowing the performance of the different batches to be compared.

Nowadays, the accelerated aging test is one of the accepted vigor tests, which aims to evaluate the ability to resist stress. Initially, the accelerated aging test was developed to estimate the longevity of seeds under storage conditions. It exposes seeds for short periods of 3 to 4 days to high relative humidity, 100% when using water or less when using saline solution, and a temperature generally in the range of 41°C, following the standard germination test. The conditions mentioned above are considered to be the most relevant environmental factors with regard to the intensity of seed deterioration, so low quality seeds deteriorate at a faster rate when compared to more vigorous ones, exposing a differentiated decrease in viability (Martins *et al.*, 2018).

To carry out the test, plastic boxes with individual compartments are used to distribute the samples (Figure 1). Inside the box there is a suspended aluminum screen, in which the seeds are distributed in a single layer, Then each box is covered and sent to the BOD (Figure 2), which is set at a temperature of 40°C to 45°C, although the recommended temperature for most crops is around 41°C. The boxes remain in the BOD for the period indicated for each crop (Martins *et al.*, 2018).

Figure 1. Plastic box with suspended aluminum mesh.

Figure 2. BOD incubator

To assess the results properly, it is advisable to measure the moisture content of the seeds before and after the accelerated ageing test. It is therefore crucial that the samples do not vary by more than 2%, as the wettest seeds are the most susceptible to the test conditions, which can lead to further deterioration (Marcos-Filho, 2015).

There are two variations of the traditional accelerated ageing test: the saline solution accelerated ageing test (SS) and the saturated saline solution accelerated ageing test (SSS), which follow the same procedure as the traditional test, differing only in that pure water is replaced by saline solutions. This allows different levels of relative humidity to be obtained. In the saline solution test, 11 milliliters of solution (11g of NaCl in 100 milliliters of water) is added to the bottom of the mini-chamber, resulting in an environment with a relative humidity of 94%. In the saturated salt solution test, 40 milliliters of NaCl (40g of NaCl in 100 milliliters of water) are used, achieving a relative humidity of 76% (Martins *et al.*, 2018).

In addition, during accelerated ageing tests, it is essential to determine the water content of the seeds and the ageing temperature to ensure uniform test conditions. Due to the less intense hydration in the SS and SSS tests compared to the traditional test, it is possible to use temperatures of up to 45°C or maintain 41°C and extend the ageing period.

The specific methodology of the accelerated ageing test varies according to the plant species and the objectives of the research, as mentioned above. However, general steps can be described:

→ Seed selection and preparation: Viable seeds free from physical damage are selected and prepared for the test, following the specific protocols for each species.

→ Accelerated Ageing: The seeds are subjected to controlled environmental stress conditions, usually combining high temperature (between 35°C and 45°C) and high relative humidity (between 90% and 95%) for a predetermined period (ranging from 24 to 96 hours).

→ Storage conditions: In some cases, seeds can be stored under sub-optimal conditions before accelerated ageing, in order to amplify the effects of the test and discriminate between lots with lower vigor.

→ Germination test: After accelerated ageing, the seeds are subjected to a germination test under ideal conditions of temperature, humidity and light, following the protocols recommended for the species.

→ Evaluation of results: Parameters such as germination percentage, germination speed index and seedling vigor are evaluated and compared between the different seed lots.

FINAL CONSIDERATIONS

The accelerated ageing test is an important tool in assessing the quality and longevity of seeds, because by subjecting seeds to controlled stress conditions, it is possible to predict their potential for storage and germination in field conditions.

The results obtained from this test generate fundamental information for producers, researchers and farmers, helping them to make decisions about the appropriate management of the seeds used. However, when analyzing the results, it is necessary to take into account the specific characteristics of each plant species, together with the environmental conditions that will be used. The accelerated ageing test is extremely important in guaranteeing the quality and success of agricultural production.

BIBLIOGRAPHICAL REFERENCES

BERTOLIN, D. C.; SÁ, M. E. de; MOREIRA, E. R. Parameters of the accelerated aging test to determine the vigor of bean seeds. **Revista Brasileira de Sementes**, v. 33, p. 104-112, 2011.

CÂMARA, J. T. da. **Factors associated with loss of quality in soybean seed production in the Planaltina-DF region**. Master's dissertation. Federal University of Pelotas. 2016.

CORRÊA, P. D. **Evaluation of the internal morphology and vigor of cotton seeds using image analysis techniques**. 2017. Dissertation (Master in Phytotechnology) - Luiz de Queiroz College of Agriculture, University of

São Paulo, Piracicaba, 2017. doi:10.11606/D.11.2018.tde-08032018-103644. Accessed on: 2024-06-16.

DUTRA, A. S.; VIEIRA, R. D. Accelerated aging as a vigor test for corn and soybean seeds. Santa Maria - RS: **Ciência Rural**. v.34, n.3, p.715-721, 2004.

FRANÇA NETO, J. B. *et al.* High quality soybean seed production technology: seed series. Londrina: Embrapa Soja, 2007. 7p. (Embrapa Soja. **Technical Circular**, 40). 2017.

MARCOS FILHO, J. Fisiologia de sementes de plantas cultivadas. 2. Ed. Londrina-PR: **Abrates**, 2015. 600p.

MARTINS, A. B. N. *et al.* Accelerated ageing test on seeds. Pelotas: **Ed. Santa Cruz**, v. 1, p. 28-32, 2018.

MICHELS, K. L. L. S. **Maturation, drying and storage of *crotalaria* seeds (*Crotalaria spectabilis Roth*)**. 2020. 86 f. Thesis (Doctorate in Agronomy) - Faculties of Agrarian Sciences, Federal University of Grande Dourados, Dourados, 2020.

RODRIGUES, L. P. S. **Effects of accelerated environmental ageing on polymer composites**. 2007. Master's dissertation. Federal University of Rio Grande do Norte. 2007.

ROSSETO, C. A. V.; FERNANDEZ, E. M.; MARCOS FILHO, J. Methodologies for adjusting the degree of humidity and the behavior of soybean seeds in the germination test. **Revista Brasileira de Sementes. Brasília**, v. 17, n.2, p. 171-179, 1995.

Chapter 3

Cold test

INTRODUCTION

The use of seeds with high physiological potential is crucial to guaranteeing productive results in economically significant crops. Good quality seeds have greater vigor, uniform germination and greater resistance to biotic and abiotic stresses, which contributes to more efficient initial establishment and more robust plants throughout the growing cycle. To ensure seed quality, an essential tool is seed analysis. This analysis involves a series of tests and evaluations that determine the viability, vigor and purity of the seeds, as well as the presence of pathogens and other impurities (MIGUEL *et al.*, 2001).

The cold test is a valuable methodology for assessing the physiological potential of seeds, especially those destined for environments with significant climatic variations. The basic principle of this test involves exposing the seeds to adverse conditions, such as low temperature, high humidity and the presence of pathogens. The soil used in the test usually comes from areas where the species is grown, which adds specific soil pathogens to the challenging conditions (Cicero & Vieira, 1994).

The cold test has been widely used in maize cultivation due to its reliable results. However, there is growing interest in standardizing it for other plant species. Although the literature points to a scarcity of research involving cotton seeds in this context, there are some studies that explore the use of the cold test to assess the physiological potential of cotton seeds (Kryzanowski,1980). In addition, there are studies investigating the use of this test in other species such as pole beans (Saminy *et al.*,1987), soybeans (Kulik *et al.*,1982; Vieira *et al.*,1992; Miguel & Cicero, 1999) and common beans (Miguel & Cicero, 1999). However, methodological studies and the application of the test have been more focused on corn seeds. This suggests a gap in research that could be explored to better understand the applicability and appropriate protocols of the cold test in a variety of plant species.

LITERATURE REVIEW

This chapter was written in such a way as to provide a theoretical approach to the conservation of recalcitrant seeds and their longevity. To gather bibliographic information, we searched for and consulted documents and information on digital academic research platforms such as scielo, periódico capes, google acadêmico, as well as scientific journal websites.

Research in the field of seed analysis has explored and suggested variations in the cold test methodology to make it more practical and accessible for laboratories. These variations include:

- Use of **Paper Towels with Soil:** First proposed by Hoppe (1956) and Crosier (1957), this methodology reduces the amount of substrate and space needed for the test, making it easier to carry out in laboratories with limited space and resources. This approach still simulates natural soil conditions, but in a more compact and manageable way.

- **Paper towel** rolls **without soil:** Loeffler *et al.* (1985) suggested a methodology that uses only paper towel rolls, without soil as a substrate. This variation is particularly interesting for determining the physiological potential of seeds for several reasons:

- **Space reduction:** The use of paper towel rolls takes up less space, allowing more samples to be tested simultaneously.

- **Elimination of the use of soil:** The absence of soil eliminates problems related to hygiene, such as the presence of pathogens and the need to sterilize the substrate.

• **Ease of Standardization:** Standardization of tests becomes easier and more consistent without the variation introduced by the use of different types of soil, resulting in more reliable and reproducible results.

These methodological variations aim to make the cold test more efficient and accessible, without compromising sensitivity and accuracy in determining the physiological potential of seeds. Adapting the methodology to reduce the use of bulky substrates and solve hygiene and standardization issues is a significant advance, especially for laboratories facing operational limitations.

Therefore, the choice of the ideal methodology will depend on the specific conditions of the laboratory, as well as the needs and objectives of the test. The variations proposed not only expand the possibilities for carrying out the cold test, but also contribute to a more efficient and practical analysis of the seeds, ensuring that high quality lots are selected for planting.

These cold tests are important for simulating adverse temperature conditions that the seeds may encounter in the field during planting, helping to determine the physiological quality of the seeds and their ability to germinate in unfavorable conditions. The results of these tests can provide valuable information for farmers when selecting the best cultivars for planting.

Different procedures for conducting the cold test have been studied, comparing it to other tests used to assess the physiological potential of cotton seeds (Miguel *et al.*, 2001).

Therefore, the methodologies used for the cold test on cotton were:

Cold test with soil - sowing in plastic boxes:

The substrate consisted of a mixture of 2/3 sand and 1/3 soil from an area cultivated with the chosen crop. This substrate was irrigated with water until it reached 60% of its retention capacity.

The seeds were subjected to different cold test conditions, varying in temperature (10°C or 15°C) and test duration (3.5 or 7 days). The conditions were as follows:

Cold test on a roll of paper with soil:

Four replicates of 100 seeds were used for each batch of each cultivar.

The Germitest paper towel rolls were moistened with water equivalent to 2.5 times the weight of the paper. The seeds were distributed over two sheets of paper and covered with a thin layer of soil from an area cultivated with beans. A third sheet of paper was then placed on top of the seeds and soil, and the rolls were made. The rolls were placed in sealed plastic boxes and kept under specific conditions that were not detailed in the description provided.

This method is commonly used to assess seed germination under controlled conditions, where the paper towel provides an ideal environment for seedling growth. The addition of soil and water to the process helps to simulate soil conditions, allowing the seeds to germinate as they would in the field. The results of this test can provide valuable information about the quality of the seeds and their ability to germinate under different climatic conditions.

These appear to be different cold tests carried out on rolls of paper with soil in temperature-controlled conditions. Each test is conducted at different temperatures (10°C and 15°C) and for different periods of time (3, 5 or 7 days). This type of test can be used to assess the resistance of certain materials, organisms or processes to low temperatures, simulating specific storage or exposure conditions.

Cold test on a roll of paper without soil:

The procedures were similar to those described in the previous section, except that the seeds were not covered with a thin layer of soil. The conditions are described below:

Due to the lack of research on chia seeds, the cold test was carried out by researchers at the Federal University of Pelotas, using moist blotting paper. The samples were exposed to different temperatures and exposure periods (6°C and 10°C for 3, 5 and 7 days).

Analyzing the germination data, it can be seen that there was no significant difference between the batches. It is important to note that when working with similar germination tests, the vigor test can detect differences in vigor (RAMOS *et al.*, 2004).

The seedling emergence test was sensitive enough to rank the lots into different quality levels, where it was possible to see that lots 4 and 6 had the highest emergence percentages, with values of 80 and 75%, respectively. Batch 2 had the lowest percentage with only 52% of seedlings emerging. When analyzing the cold test data, it could be seen that it was not efficient enough to rank the batches satisfactorily in terms of vigor. For both temperatures (10 and 6 °C), it was only on the third day of exposure that the batches were separated into different levels, although this is not entirely in line with the data obtained from the seedling test and the emergence speed index. The other exposure times (5 and 7 days) for both temperatures (10 and 6 °C) did not reveal any differences in terms of vigor. It was therefore concluded that the cold test was not efficient for assessing the vigor of chia seeds.

FINAL CONSIDERATIONS

The seed cold test is an essential tool for guaranteeing the quality and viability of seeds in low temperature conditions. Carrying out these tests properly allows farmers and producers to select more resistant varieties, better plan their harvests and ensure the viability of seeds during storage. Compliance with standards and regulations is crucial to obtaining valuable and reliable results, contributing to sustainability.

BIBLIOGRAPHICAL REFERENCES

MIGUEL, M.H., CARVALHO, M.V., BECKET, O.P., FILHO, J.M. Cold test to evaluate the physiological potential of cotton seeds. **Scientia Agricola**, v.58, n.4, p.741-746, 2001.

*MIGUEL, M.H., CICERO, S.M. Cold test in the evaluation of bean seed vigor. **Scientia Agricola**, v.56, n.4, p.1233-1243, 1999.

MAASS, D.W ; PIEPER, M.S; HELING, V.L; MENEGHELLO, G.E.Standardization of the cold test methodology for evaluating the vigor of chia seeds. Anais Congresso de iniciação científica, **UFPel**, 2017.

Chapter 4

Tetrazolium test

INTRODUCTION

According to Delouche *et al.* (1976), the tetrazolium test is based on the activity of dehydrogenase enzymes in tissue respiration processes. During respiration, hydrogen ions are released, with which the 2,3,5 triphenyl chloride salt of tetrazolium reacts to form a red-colored, insoluble substance called formazan in the living tissues of the seed.

Brasil (2009) points out that it is a test used to assess the physiological quality of seeds. The aim of this test is to determine viability, which is an estimate of seed germination, particularly of species that germinate slowly in normal tests or that do not germinate when subjected to commonly used methods because they are dormant.

According to Piña-Rodrigues & Santos (1988); França Neto (1999), in the seed quality control system, the tetrazolium test is used to assess the physiological quality of agricultural seeds, such as: soybeans, wheat, corn, cotton, beans and forage plants of the Brachiaria genus, and has taken on a prominent position for some crops, due to the large amount of information provided by the test (França Neto, 1999).

Brasil (1992) explains that the tetrazolium test is also used to quickly determine the viability of seeds, particularly species that germinate slowly in normal tests or that do not germinate when subjected to commonly used methods because they are dormant.

It has been applied as an alternative to the germination test to assess the viability of seeds of Bertholletia excelsa (Camargo *et al.*, 1997), Heteropteris aphrodisiaca (Arruda, 2001), Cordia trichotoma (Mendonça *et al.*, 2001), Astronium graveolens, Jacaranda cuspidifolia and Parapiptadenia rígida (Fogaça, 2003), among other species.

The test consists of assessing the quality of seeds based on the change in color of the tissues in the presence of a solution of tetrazolium salt (2, 3, 5 -

triphenyl tetrazolium chloride, or TCT), which is reduced by the enzymes of living tissues, resulting in a compound called formazan, colored carmine red (Vieira & Von Pinho, 1999). According to Delouche *et al.* (1976), the reaction allows a very clear delineation between breathing tissue, which acquires a red color, and dead or decayed tissue, which retains its original color. The color pattern of the tissues can be used to identify viable and non-viable seeds.

According to Marcos Filho *et al.* (1987); Piña-Rodrigues & Santos (1988), the reaction of tetrazolium in the seed is affected by various factors such as preconditioning and cutting of the seed, temperature and incubation time, pH and concentration of the solution, age and deterioration of the seed. The seeds are subjected to prior preparation such as preconditioning and cutting or removal of the integuments, which must be carried out carefully so as not to interfere with the results obtained. It is recommended that the temperature at which the coloration develops is between 35 and 40°C, as this is faster, but the test can normally be carried out at temperatures between 20 and 40°C. The pH of the solution should be neutral, around 6.5 to 7.0, since more acidic or basic solutions alter the speed and intensity of tissue staining, making interpretation difficult.

The tetrazolium test enables a rapid assessment of the probability of the seeds, making it possible to obtain results in a few hours, unlike other tests that can take days or weeks.(Seed Viability and Vigor Tests" in Seed Technology and Its Biological Basis 2001).As well as indicating predictions, the tetrazolium test also provides information on the physiological quality of the seeds, identifying physical and physiological damage.(Krzyzanowski, FC, *et al.* "Tetrazolium test for seed viability" in Seed Vigor Testing Handbook 1999).

The test is effective in identifying various types of damage, including mechanical damage, damage caused by and damage caused by pathogens. (Moore, RP "History Supporting Tetrazolium Seed Testing" in Seed Technology 1985).

The tetrazolium test is applicable to a wide range of seed species, both agricultural and forestry. (AOSA (Association of Official Seed Analysts). "Tetrazolium test manual" 2010).

MATERIAL AND METHODS

Step-by-step tetrazolium test:

1. Sample Collection and Preparation:

- Sample selection: Choose a representative sample of the seeds to be tested.

- Seed cleaning: Removes impurities and residues from the seeds.

2. Moisturizing the seeds:

- Soaking in water: Place the seeds in distilled water for 12 to 18 hours.

- Temperature: Keep the water temperature between 20°C and 30°C.

3. Preparing the seeds for the dye:

- Cut the seeds: Make cuts that allow the dye to penetrate (longitudinal or transversal, depending on the species).

4. Tetrazolium immersion:

- Solution preparation: Dissolve 1g of tetrazolium in 1 liter of distilled water to obtain a 0.1% solution.

- Submerge the seeds: Place the cut seeds in the solution.

- Temperature and time: Keep the solution between 30°C and 40°C in a dark environment for 1 to 4 hours.

5. Washing the seeds:

- Removing the dye: Rinse the seeds under running water after processing.

6. Analysis of results:

- Observation of coloring: Examine the seeds under a magnifying glass.

- Interpretation: Viable parts turn red or pink; non-viable parts remain colorless.

7. Classification and reporting:

- Classification: Categorize seeds into viable and non-viable.

- Report: Document the results

Alternatives:

❖ Substitution of distilled water: Use filtered and boiled water.

❖ Temperature control:

o Home Thermometer: Household uses.

o Improvised water bath: Use containers with hot water to maintain the temperature.

❖ Cutting blades: Stylish or new razor blades.

❖ Precise Measurement: Domestic precision scales for measuring tetrazolium.

❖ Dark environment: Cardboard boxes or dark cupboards.

IMAGES

Photos: José de Barros França-Neto

Photos: José de Barros França-Neto

Photos: José de Barros França-Neto

Photos: José de Barros França-Neto

Photos: José de Barros França-Neto

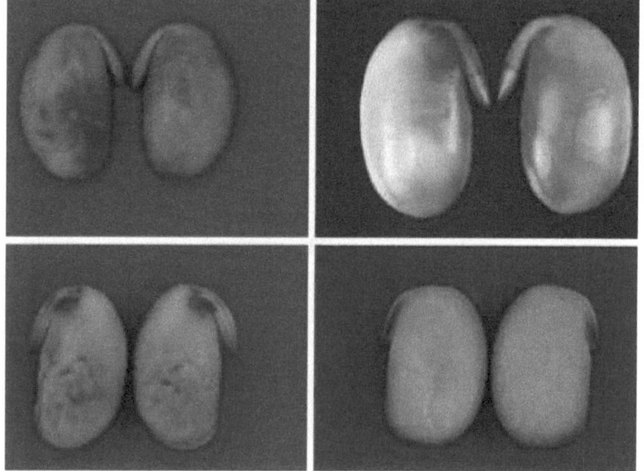

Photos: José de Barros França-Neto

FINAL CONSIDERATIONS

The tetrazolium test is a crucial method for assessing the physiological quality and accuracy of seeds, providing quick and precise results. It stands out for its ability to identify physical and physiological damage that can impair germination, especially in species that are dormant or germinate slowly.

The thorough test, which includes hydration, cutting, staining and analysis procedures, guarantees accurate identification of viable sensations. The test is used on a variety of agricultural and forestry species, making it crucial for seed quality control. As such, it is an effective alternative to seed testing.

BIBLIOGRAPHICAL REFERENCES

ARRUDA, J.B. **Aspects of the germination and cultivation of the dog knot (*Heteropteris aphrodisiaca O. Mach.*).** 2001. 142p. Dissertation (Master's Degree in Tropical Agriculture) - Faculty of Agronomy and Veterinary Medicine, Federal University of Mato Grosso, Cuiabá.

BRAZIL. Ministry of Agriculture and Agrarian Reform. **Rules for seed analysis.** Brasília: SNAD/CLAV, 1992. 365p.

BRAZIL. Ministry of Agriculture and Agrarian Reform. **Rules for seed analysis**. Brasília: CLAV/DNDV/SNAD, 2009. 365p.

CAMARGO, I.P.; CARVALHO, M.L.M.; VIEIRA, M.G.G.C. Evaluation of deterioration in Brazil nut seeds by the tetrazolium test. **Pesquisa Agropecuária Brasileira**, v.38, n.8, p.835-9, 1997.

DELOUCHE, J. C.; STILL, T. W.; RASPET, M.; LIENHARD; M. **The tetrazolium test for seed viability**. Brasília: AGIPLAN, 1976.

FOGAÇA, C.A. **Use of the tetrazolium test to assess the viability of forest seeds**. 2003. 50p. Dissertation (Master's Degree in Agronomy) - Faculty of Agricultural and Veterinary Sciences, Universidade Estadual Paulista, Jaboticabal.

FRANÇA NETO, J.B. **Tetrazolium tests for determining seed vigor**. In: KRZYZANOWSKI, F.C.; VIEIRA, R.D.; FRANÇA-NETO, J.B. (Ed.). **Seed vigor**: concepts and tests. Londrina: ABRATES, 1999. p.8.1-8.5.

MARCOS FILHO, J.; CÍCERO, S.M.; SILVA, W.R. **Evaluation of seed quality**. Piracicaba: FEALQ, 1987. 230p.

MENDONÇA, E.A.F.; RAMOS, N.P.; PAULA, R.C. **Viability of *Cordia trichotoma* (Vellozo) *Arrabida ex Steudel* (black laurel) seeds by the tetrazolium test. Revista Brasileira de Sementes**, v.23, n.2, p.64-71, 2001.

PIÑA-RODRIGUES, F.C.M.; SANTOS, N.R.F. Tetrazolium test. In: PIÑA-RODRIGUES, F.C.M. (Coord.). **Manual de análise de sementes florestais**. Campinas: Cargill Foundation, 1988. p.91-100.

VIEIRA, M.G.G.C.; VON PINHO, E.V.R. **Methodology of the tetrazolium test on cotton seeds**. In: KRZYZANOWSKI, F.C.; VIEIRA, R.D.; FRANÇA-NETO, J.B. (Eds.). **Seed vigor**: concepts and tests. Londrina: ABRATES, 1999. p.8.1.- 8.12.Castro, Y. G. P.; Krug, H. P. (1951). **Experiments on the germination and conservation of seeds of *Inga edulis*, a species used for shading coffee trees**. Ciência e Cultura, 3: 263-264.

COSTA, C. J. (2009). **Storage and conservation of seeds from Cerrado species**. Embrapa Cerrados-Documents (INFOTECA-E).

ELLIS, R. H. *et al.*(1990). **Low moisture contente limits to relations between seed longevity and moisture**. Annals of Botany, 65:493-504.

FAIAD, M. G. R. *et al.* (2005). **Strategies and Results of Long-Term Seed Germplasm Conservation**. Available at: . Accessed on Aug 27, 2020.

FERREIRA, S. A. do N.; GENTIL, D. F. de O. (2003) **Storage of camu-camu seeds (*Myrciaria dubia*) at different humidity levels and temperatures. Revista Brasileira de Fruticultura**, v. 25, n. 3, p. 440-442.

FONSECA, S. C. L., FREIRE, H. B. (2003). **Recalcitrant seeds: post-harvest problems**. Bragantia, 62: 297-303.

GARCIA, C. *et al.*(2014). **Preservation of viability and vigor of *Araucaria angustifolia* (Bertol.) Kuntze seeds during storage**. Ciência Florestal, 24 (4): 857-867.

GENTIL, D. F. O. (2003). Conservation of *Myrciaria dubia* (H.B.K.) McVaugh seeds. Thesis (Doctorate in Agronomy) - **Luiz de Queiroz College of Agriculture**, Piracicaba, 41f.

FRANCO, D. F. *et al.*(2016). **Seed storage**. Embrapa Clima Temperado-Technical Communication (INFOTECA-E).

JONES, H. A. (1920). **Physiological study of maple seed. Botanical Gazette**, 69:127-152.

KIDD, F. (1914). **The controlling influence of carbon dioxide on the ripening, dormancy and germination of seeds**. Part II. Proceedings of the Royal Society of London, 87:609-625.

MARTINS, C. C. *et al.* (2009). **Drying and storage of juçara seeds. Revista Árvore**, 33: 635-642.

MEDEIROS, A. C. de S., & EIRA, M. T. S. (2006). **Physiological behavior, drying and storage of native forest seeds**. Embrapa Florestas-Technical Circular (INFOTECA-E). Available at: https://ainfo.cnptia.embrapa.br/digital/bitstream/CNPF-2009-09/41479/1/circ-tec127.pdf

NASCIMENTO, W. M. O. do *et al.* (2010). Conservation of açaí seeds (*Euterpe oleracea Mart.*). **Revista Brasileira de Sementes**, 32(1): 24-33.

NEVES, C. S. V. J. (1994). **Recalcitrant seeds: a literature review**. Pesquisa Agropecuária Brasileira, 29(9): 1459-1467.

PRANGE, P. W. (1964). **Conservation study of the germination power of *Araucaria angustifolia* (Bert.) Oktze seeds**. Anuário Brasileiro de Economia Florestal ,16 :43-48.

ROCHA, M. do S. *et al.* (2010). **Cryopreservation of jatropha zygotic embryonic axes**. In: Congresso Brasileiro de Mamona, 4.; Simpósio Internacional de Oleaginos Energéticas, Inclusão social e energia: annais. Campina Grande: Embrapa Cotton.

VIEIRA, F. de A.; GUSMÃO, E. (2008). **Biometry, seed storage and seedling emergence of *Talisia esculenta Radlk.* (Sapindaceae)**. Ciência e agrotecnologia, 32 (4):1073-1079.

WALTERS, CHRISTINA *et al.* (2013). **Preservation of recalcitrant seeds**. Science, 339(6122): 915-916.

ZINK, E.; ROCHELLE, L. A. (1964).**Estudos sobre a conservação de sementes**. XI - Cocoa. Bragantia, 23: 111-116.

Chapter 5

Emergency test

INTRODUCTION

The emergence test is a seed quality assessment method that simulates adverse or stressful field conditions to measure the seeds' ability to germinate and emerge. Unlike the germination test, which is carried out under ideal and controlled conditions, the emergence test exposes the seeds to stress factors such as temperature (Avila *et al.*, 2019), salinity (Pagliarini *et al.*, 2022) or humidity (Zhang, Liu and Chen., 2012). This test is essential for predicting the actual performance of seeds when planted in environments where they will face natural challenges.

Seed quality is a crucial factor in agriculture, directly impacting productivity and crop success. The emergence test is vital for several reasons. Firstly, it offers a more realistic assessment of how seeds will behave in field conditions, which are often less favorable than ideal laboratory conditions. It helps to identify seeds that have the ability to germinate and grow in adverse conditions, which is crucial for crop success in regions with challenging climates or poor soils. In addition, the emergence test measures not only the germination capacity, but also the vigor of the emerging seedlings. Vigorous seeds produce more robust seedlings (Bagateli *et al.*, 2020), capable of competing with weeds and resisting pests and diseases. The general assessment and development of seedlings, including the length of roots and stems, is an indicator of the plants' ability to establish themselves well in the field.

The emergence test is also fundamental to breeding programs, helping to select seed varieties that are more resilient to stressful conditions, resulting in more adaptable and productive cultivars. It provides valuable data for farmers, allowing for more informed decisions about which seed lots to plant in different types of soil and climatic conditions. In addition, by identifying low-quality seeds before planting, the emergence test helps to minimize losses in productivity and resources by avoiding sowing seeds that will not thrive (Bagateli *et al.*, 2020).

This test also allows for better agricultural planning and management, helping to mitigate risks in agricultural production.

MATERIALS AND METHODS

In order to prepare this chapter of the literature review on the seed emergence test, a systematic approach was adopted to ensure the comprehensiveness and relevance of the information collected. Firstly, the specific objectives of the review were defined, focusing on the importance of the emergence test, its practical applications, methodologies employed and impacts in different agricultural contexts. Search terms such as "seed emergence test", "germination under stress" and "climatic adaptation of seeds" were identified to guide the search for relevant literature. The searches were carried out in renowned scientific databases such as Scopus, Web of Science, PubMed, Google Scholar and SciELO, as well as specific agronomy and botany publications.

RESULTS AND DISCUSSION

TEST OBJECTIVES

The emergence test assesses the ability of seeds to germinate and develop into seedlings under less ideal or more stressful conditions than those used in the standard germination test. It simulates more rigorous field conditions to provide a more realistic idea of how seeds will behave in practical situations.

This test can be carried out in the laboratory or in the field. In the laboratory, controlled stress conditions are tested, such as temperature, salinity and humidity, similar to those experienced by the seeds in the field. In the field test, however, the seeds are subjected to real environmental conditions and are influenced by all aspects of the location.

IMPORTANCE OF THE EMERGENCY TEST

The emergence test plays a crucial role in assessing seed quality, offering information beyond what the standard germination test can provide. While the germination test is carried out under ideal and controlled conditions, focusing mainly on the basic viability of the seeds, the emergence test simulates adverse and stressful field conditions. This more rigorous approach allows for a realistic comparison of how seeds perform in practical situations, helping to predict how they will behave when exposed to the natural challenges found in different agricultural regions (Munyaneza *et al.*, 2022).

In field conditions, seeds often face factors that are not replicated in standard germination tests (Oregon State University). The emergence test, by exposing seeds to sub-optimal conditions, allows farmers to identify those that have the greatest capacity to germinate and develop even in adverse environments. This is particularly important in regions of difficult agricultural production, where the success of crops depends on the resistance and vigor of the seeds. By providing a more practical and accurate assessment of seed potential, the emergence test helps to choose lots that offer greater security and predictability when planting, resulting in better harvests and greater productivity.

In addition, the emergence test offers significant advantages for breeding programs. The results of this test are essential for selecting and developing seed varieties that are more robust and adaptable to stressful conditions (Martins *et al.*, 2016). The ability to identify seeds with greater vigor and resistance allows breeders to focus their efforts on genotypes that possess desirable characteristics, such as drought tolerance (Filho *et al.*, 2023), disease resistance and nutrient use efficiency. This not only contributes to the development of more resilient and productive cultivars, but also strengthens agricultural sustainability, since more robust varieties can reduce the need for agricultural inputs and increase crop stability.

EMERGENCY TEST IN THE FACE OF CLIMATE CHANGE

Climate change is drastically altering weather patterns, exposing seeds to various abiotic stresses such as drought, salinity, flooding and extreme temperatures. These conditions can compromise germination (Gya *et al.*, 2023) and seedling vigor (Demir *et al.*, 2022), and can thus lead to a reduction in crop yields, putting the food security of populations around the world at risk.

By assessing seed performance under different stress scenarios, the Emergence Test provides valuable information for farmers, researchers and decision-makers. Through this testing, it is possible to identify seed varieties that are more resilient to climate change, ensuring that agriculture adapts to new challenges and that food is produced even under conditions of stress, especially thermal, water and salt stress, which are the most impactful aspects on seeds and the most studied today (Sisodia and Sharma, 2023). But the importance of seed testing is not limited to agriculture. The seed industry also benefits from this process, as it allows seed quality to be assessed and batches with low viability to be identified. This ensures that farmers receive high-quality seeds, increasing the chances of successful planting and crop productivity.

Therefore, the seed emergence test is crucial in the face of climate change, as it provides a realistic assessment of the ability of seeds to germinate and grow in adverse conditions. By simulating the environmental stresses that plants face in the field, this test identifies the most vigorous and resilient seeds, which are essential for agricultural sustainability. As well as helping to select adaptable cultivars, the emergence test minimizes losses, saving valuable resources and ensuring continuous and stable production. As such, it is an indispensable tool for tackling climate challenges and ensuring global food security in the future.

FINAL CONSIDERATIONS

The emergence test is an essential tool for assessing seed quality, offering an objective view of seed performance under field conditions, evaluating seedling vigor, helping to select resilient varieties and contributing to risk reduction in agriculture.

BIBLIOGRAPHICAL REFERENCES

ÁVILA, M. R. *et al*. Dynamics of the rangpur lime seed germination test conducted under different temperatures. **Journal of Seed Science**, v. 41, n. 3, p. 344-351, jul. 2019.

BAGATELI, J. R. *et al. Seed* vigor and population density: effects on plant morphology and soybean productivity. **Brazilian Journal of Development**, v. 6, n. 6, p. 38686-38718, 18 jun. 2020.

DEMIR, I. *et al*. Radicle Emergence as Seed Vigour Test Estimates Seedling Quality of Hybrid Cucumber (*Cucumis sativus* L.) Cultivars in Low Temperature and Salt Stress Conditions. **Horticulturae**, v. 9, n. 1, p. 3, 1 Jan. 2023.

FILHO, C. C. F. *et al*. Breeding for drought tolerance in perennial ryegrass (*Lolium perenne* L.) and tall fescue (*Lolium arundinaceum* [Schreb.] Darbysh.) by exploring genotype by environment by management interactions. **Grassland research**, v. 2, n. 1, p. 22-36, 1 mar. 2023.

GYA, R. *et al*. A test of local adaptation to drought in germination and seedling traits in populations of two alpine forbs across a 2000 mm/year precipitation gradient. **Ecology and Evolution**, v. 13, n. 2, p. 1-19, Feb. 1, 2023.

MARTINS, C. C. *et al.* Methodology for the selection of soybean strains for germination, vigour and field emergence. **Revista Ciência Agronômica**, v. 47, n. 3, p. 455-461, jul. 2016.

MUNYANEZA, V. *et al.* Various vigour test methods to rank seed lot quality and predict field emergence in two forage grasses. **Seed Science and Technology**, v. 50, n. 3, p. 345-356, 2022.

PAGLIARINI, M. K. *et al.* Seeds germination of different species in saline water. **Research, Society and Development**, v. 11, n. 11, p. e115111133135-e115111133135, 17 aug. 2022.

SEED LABORATORY. **Importance of Seed Vigor Testing**. Available at: <https://seedlab.oregonstate.edu/importance-seed-vigor-testing>. Accessed on: June 23, 2024.

SISODIA, R; SHARMA, R. Bibliometric Analysis of Peer-Reviewed Literature on Stress Factors Affecting Agricultural Productivity. **Current Agriculture Research Journal**, v. 10, n. 3, p. 170-180, Jan. 5, 2023.

ZHANG, D.; LIU, X.; CHEN, F. Q. Effects of Temperature and Humidity on the Germination of Two Pioneer Species in Ecological Restoration. **Applied Mechanics and Materials**, v. 209-211, p. 1265-1268, Oct. 2012.

Andréa Bicca Noguez Martins - Agronomist, Professor IFSul Campus Bagé, PhD in Seed Science and Technology and Post-Doctoral Student in Plant Health UFPel/Pelotas.

Andressa Alves Cassão- Undergraduate student in Agronomic Engineering IFSul Campus Bagé.

Andressa Meneses Prockt- Undergraduate student in Agronomic Engineering IFSul Campus Bagé

Bruna Franco da Silva- Undergraduate student in Agronomic Engineering IFSul Campus Bagé

Daniele Pacheco da Silva- Undergraduate student in Agronomic Engineering IFSul Campus Bagé

Gabriele Padilha Schneider- Undergraduate student in Agronomic Engineering IFSul Campus Bagé

Joice Fernanda Lübke Bonow- Agronomist, Professor IFSul Campus Bagé, PhD in Plant Health UFPel/Pelotas.

Luis Filipe Oliveira de Leon- Undergraduate student in Agronomic Engineering IFSul Campus Bagé

Paula Cilene Machado Munhoz- Undergraduate student in Agronomic Engineering IFSul Campus Bagé

Pierri Caldas Bauerman- Undergraduate student in Agronomic Engineering IFSul Campus Bagé

Willian Siqueira da Silva- Undergraduate student in Agronomic Engineering IFSul Campus Bagé

Ygor Mota Soca Machado- Undergraduate student in Agronomic Engineering IFSul Campus Bagé

I want morebooks!

Buy your books fast and straightforward online - at one of world's fastest growing online book stores! Environmentally sound due to Print-on-Demand technologies.

Buy your books online at
www.morebooks.shop

Kaufen Sie Ihre Bücher schnell und unkompliziert online – auf einer der am schnellsten wachsenden Buchhandelsplattformen weltweit! Dank Print-On-Demand umwelt- und ressourcenschonend produziert.

Bücher schneller online kaufen
www.morebooks.shop

Printed by Books on Demand GmbH, Norderstedt / Germany